THE WEEKEND GOLD MINER

By A.H. Ryan, Ph.D.

Published by:

Gem Guides Book Co.

315 Cloverleaf Drive, Suite F
Baldwin Park, CA 91706

Table of Contents

Foreward to the Third Edition

When the first edition of the weekend gold miner was written, the price of gold was pegged at around thirty dollars a troy ounce. Today it is selling at more than a dozen times that amount. This adds a little zest to the hobby of hunting for gold.

There is still gold waiting to be found. A friend of mine has a favorite "diggin' place" up in the central mountains of Arizona; he's gone there almost every weekend of every summer for the eight years I've known him. He brings back a little gold each time.

By now he has quite a little pile of "dust" and nuggets – his retirement nest-egg. And he has the best tan you ever saw, and the strongest shoulder muscles. Yep, pardner, there's more than gold out there!

Of course, the high value of gold changes the picture for the weekender. Many locations that were once unclaimed or abandoned are now claimed or even being worked. So, you need to take more care to avoid trespass violations. Still, there are many places to work and gold to be found.

CHAPTER 1

Twentieth Century Sourdoughs

Tickling the western flank of the Sierra Nevada range, the Yuba River gurgles its way toward the ocean through a stand of cool, beautiful forest, not far from towns (living and dead) with such names as Rough and Ready, Placerville, and Gold Run. This is Forty-niner country. And it is trout country. It was trout I was after, one weekend a few years ago – trout and scenery. Gold was only a dim legend to me.

We swung the old car into a side road not far from the river, and wallowed down the ruts to a grassy camping ground near the bank. There we pitched our tent in a little triangle of trees, close enough to the shallow rapids so the sound would sing us to sleep at night. My son and I put up the tent while mother explored the campground. There were only a few others staying there – a camper/truck combination, a small trailer, and a trio of pup tents next to a battered old Dodge. On her return my wife reported the camper was occupied by a nice middle aged lady whose husband was off fishing, the trailer by a young couple, and the pup tents by some young college boys.

The fish were calling so as soon as the tent was up we menfolk headed downstream with the fishing tackle, leaving the wife to get the place organized. We did pretty well, and in late afternoon trudged back with half a dozen nice trout in our creels. On the way we met the college boys. They were wearing wet-suits because the water was rather chilly, and masks with long snorkel tubes on them. Two of the boys were floundering through the rapids, towing a strange sort of home made raft, while the third was standing near the bank looking at something in a little plastic bottle.

He glanced up as we passed, and grinned. "How'd you make out? Catch anything?"

"We did all right." I opened the creel to show him our fish.

"Nice." He looked at our catch and nodded appreciatively. "We had a good day, too."

He held up the little plastic bottle. A quiver went through me as I saw the soft yellow sheen of the pebbles inside. Was it...? No, of course not.

I blinked at him. "That isn't....? I mean, it's not....?"

He grinned. "Sure it is. Gold nuggets. Here, look."

He dropped the bottle into my hand, which bobbled with the unexpected weight. Gold, native gold, taken from the deposits where it lay. Then and there I was bitten by the gold bug, and the fever has never gone down.

The fellows were diving for gold in the river, scooping up sand and gravel from the upstream side of the rocks where the weight of the nuggets was too much for the current to dislodge them. Earlier in the year the swift surge of the spring freshets had torn sand, soil and gold from the banks high up in the mountains and swept the mixture down to where the current was weaker. There the heavy gold had settled to the bottom and bobbled slowly downstream until the rocks in the rapids caught and held nuggets.

The boys used the raft as a working platform, diving to scoop up bucketsful of gravel from under the rocks and then dumping the stuff on the raft to sort at their leisure. They were interested only in nuggets, and a sieve was the only separation equipment they needed. The three of them had been there about a week, and already they had collected more than four hundred dollars' worth of gold – plenty to cover the expenses of their vacation.

"What happens to the gold dust?" I asked innocently. In all the movies I'd seen, the old prospector shows up at the Longbranch Saloon with a little buckskin bag of gold dust.

He shrugged. "It's carried on down by the current until it settles out on the bank or behind some other rock."

I glanced at the sand under my feet. "Is there gold here, do you suppose?"

"Probably. I'll lend you a pan, if you want to try it."

Did I! In less than an hour, the three of us – my wife, my son, and I – were squatting by the river, panning gold. At first I wielded the gold pan while my son and my wife inspected the operation a bit skeptically. But after the first few traces of color in the pan, they insisted on trying it, and I ended up with the shovel. We didn't get much. But it was enough to cover the bottom of the plastic medicine bottle, and when we got home, we displayed it triumphantly to all our friends.

The old Mother Lode country where we were camping is still the best place for weekend prospecting. But it's not the only place. In the winter months, when the northern hills are too cold for comfort, we've turned to the desert for our recreation. From about October to May, the Mojave Desert is usually quite comfortable in the daytime, though a bit chilly at night. During this season, the desert loses its man-killing ferocity, although one must never make the mistake of treating the place lightly. Like any acquaintance of strong character, the desert becomes either a close friend or a hated enemy. For my part, I like it, especially in the cool months. Then the occasional showers come, and the desert responds with a hopeful, almost pitiful show of green in the creek beds. And in the spring the stark, fantastic Joshua trees and yuccas put out weird blossoms in great profusion. And always the desert is vast and silent and smog-free.

The desert has gold. The gold rush hit there later than the stampede into central California, and it attracted much less attention. For one thing, mining was more rugged in the desert, and those who came for easy pickings soon moved on. For another, the shortage of water in the desert made transportation the key factor in the economics of mining. Nevertheless, a few miners struck it rich, skimmed off the cream, and went elsewhere, leaving the hills dotted with abandoned shafts. Leaving legends, too, of lost mines with fabulous riches. In a later chapter we'll pass on some of these legends to you.

Weekend gold miners fall into two types. One group is really after souvenirs, like the little sprinkling of gold we got on our first try, or an occasional little nugget. These treasures can be mounted to wear as jewelry, or potted in clear plastic as conversations pieces. These souvenir hunters usually work with small gold pans by the side of a stream, or with dry or "water-saving" equipment if they're in the desert. With reasonable luck, they expect to get a few dollars' worth of dust or a little nugget for a weekend's work. Sometimes they find other treasures, like old bottles, coins, or other mementos. And maybe they fish or watch the scenery or just loaf, in between spells of panning dirt.

The other type of amateur gold miner takes it a little more seriously – and works a little harder. He may have found a particularly rich place on an earlier trip, and gone back to work it week after week. This type usually finds it worthwhile to buy or build a good-sized sluice or rocker, with which they can process a lot of dirt over a weekend. Two men, working a rich strike, can recover several hundred dollars' worth of gold over a weekend, with the proper equipment and good strong backs. Not bad, not bad at all. Why don't they become professional and do it full time, you may ask, at four or five hundred dollars a day. The reason is that professional gold mining is a lot like professional gambling. You may be riding high one day, the envy of all your friends; then the lode, or your luck, runs out. You're broke again.

And all gold miners are dreaming – whether they admit it or not – of the bonanza, the rich pocket of gold that can bring a small fortune. A few years ago an anonymous correspondent reported in *Desert Magazine* that he had found the location of one of the lost gold mines of the Mojave – Pegleg's Mine. He never revealed where he found it – for very good reasons – but he did substantiate his statements by sending in some gold nuggets, an old belt buckle, and part of an old Spanish sword-scabbard. He claimed to have taken $3,000,000 from the mine. Whether he was telling the truth remains a secret between him and the Internal Revenue Service. Maybe he was spoofing us, but I don't think so. It would seem to be a very pointless joke, and the buckle and scabbard, along with the nuggets, were certainly authentic. And we know the gold was in the mine in years past.

Maybe next time you'll be the lucky one.

CHAPTER 2

Where to Look

The old saying of the 49ers is still valid – gold is where you find it. Of course, there's a better chance of finding it in some places than in others, but there are plenty of examples of ignorant prospectors chasing their burros into unlikely areas and striking it rich. Maybe their burros tipped them off. Burros are smart.

However, it turns out that the ignorant prospector was usually in the right general area; if the prospector had been traveling with a trained geologist instead of a burro, the geologist would have recommended the same area that the burro did. There are certain signposts and clues in the geology and rock formations of an area that point to the presence--now or in the past--of gold in practical quantities.

Gold is one of the basic elements of which the universe is composed, and is found everywhere in more or less abundance. Even sea water contains gold – one pound per 5000 tons of water. Not enough to be practically useful until some bright inventor comes up with an economical way to get it out. If he does he'll get rich, because there's an awful lot of water in the oceans. Someone calculated that these traces of gold in sea water come to a total of ten billion tons of gold. So it's worth thinking about.

For most of us, though, it is much easier to look for gold on land. Everyone has heard about the gold rush of the 1840s and 50s that took gold seekers to California and then to Colorado. But gold has been found in significant amounts in at least 21 states of the United States. Native Americans were using gold long before there was any recorded history of mining. They found it wherever mountain streams or rivers washed away old mountains.

Since Thomas Jefferson first described gold findings in the Rappahannock River in 1782 and gold was discovered in North Carolina in 1793 (a seventeen pound nugget was found in this state in 1799), gold has been found and mined in all periods of our history. It was found in Alaska in the 1790s, but fortunately for the United States, did not raise any interest until after the Russians sold the land. In the 1820s, Georgia had its own gold rush; enough gold was found that the state had its own mint at Dahlonga. Glacial deposits left substantial gold deposits in Indiana and Ohio. Gold discovered in Alabama in 1830 was mined only lightly until the price of gold made exploration worthwhile, then resulted in a small gold rush from 1930-1941. Pennsylvanians collected 40,000 ounces of gold while mining iron. Maine, with discoveries in 5 counties in 1985 and Connecticut, where traces of gold were found in the Meshomiac State Forest in 1984, are the newest state entries into the gold business.

Gold is often found in operations where other metals are mined. Utah reported the discovery of twenty thousand ounces, mostly from copper mining in Bingham City. If you are looking for the metal in West Virginia or Indiana keep an eye out for diamonds. Twenty or more of these gems were found in the heyday of gold hunting just before the 1900s.

Although gold has been found in almost every state, it is more plentiful in some than in others. The western states, the Black Hills of South Dakota, and Alaska have been the biggest producers. Trying to specify where to look within that area is difficult. It would be nice to have a treasure book to tell you exactly where to find that rich strike, but it wouldn't be very practical, because everyone would descend on that rich strike and it wouldn't remain rich very long. The most any guide can do is indicate where gold has been found in the past. From that point on it's a matter of observing the clues and following the signposts--and having a little luck. The odds on finding a fortune are pretty slim, but there's an excellent chance of finding at least some gold.

In my opinion the five best bets for the aspiring placer-miner are California, Arizona, Colorado, Nevada, and South Dakota. In each of these states, some areas will be well worth exploring, while others will be barren. In order to simplify the discussion, I have selected what I consider to be the most promising areas in each state, and described the locations in those areas where gold has previously been found.

In what follows, only general locations are shown on the maps, with more detailed information in the text. They should be used in conjunction with a good state map as least as detailed as the Rand McNally road maps.

California

1. State Highway 49 area. This charming little road, running north and south from about Sierra City northeast of San Francisco to around Angel's Camp due east of San Francisco, was one of the favorite hunting grounds of the old '49 prospectors--hence the name. Gold has been mined there for 150 years, and there's more left. Good bets along 49 are Fiddletown (in Amador County); Calaveras, Camache and Murphy's in Calaveras County, and Auburn and Loomis, along the American River in Placer County.

2. Humboldt County in Northern California, along the coast. There has been placer mining along the Klamath and Trinity Rivers, and beach gold has been found where the rivers and smaller streams join the ocean. The beach gold is usually in the form of fine flakes, requiring skillful panning. Hard work, but the scenery makes up for it.

3. Kern County, inland from Santa Barbara, has been a big producer of gold, but most of it has been hard-rock (lode) mining, not for the casual hunter.

Nevada

Placer mining in Nevada, the fifth largest producer of gold, has been done mainly along the southwestern boundary and along the northern boundary of the state. In addition to gold, the state has produced a great deal of silver. Good bets are:

1. Eureka County, especially northwest of Carlin, along Lynn Creek.

2. Esmeralda County, which started out as a producer of silver, had a fairly intense burst of gold production around 1905. Gold Mountain just north of tiny State Route 71 in the southern part of the county, might be fun to try if you don't mind roughing it.

3. Elko County along the Bruneau and the Owyhee Rivers, especially near the head waters of the Owyhee. These are about 50 miles east of Midas, and 45 miles northwest of Elko, over rough terrian.

4. Humboldt County. Try all creek beds, old terraces, etc. Explore the bed of Pole Creek, about 45 miles north of Winnemucca on US 95.

5. Lynn County. While you're on your way to or from Virginia City (a fascinating Old West town reincarnation) try the creeks around Silver City, five miles south.

Arizona

Gold is distributed along a broad line going from northwest Arizona to the southeast corner of the state.

1. Around Clifton, in Greenlee County. Look along the San Francisco River, Eagle Creek and Chase Creek.

2. Maricopa County. Near Wickenberg, try along the Hassayampa River where the old Vulture Mine has reopened as a tourist attraction to show what a gold mine looks like.

3. In Mohave County, the Mohave Mountains, eighteen miles southeast of Topock. This is where the Red Hill placer diggings, Printer's Gulch and Dry Gulch were.
4. The Gold Basin area is also in Mohave County, near Chloride, about 20 miles north of Kingman.
5. The Superstition Mountains, east of Apache Junction near Phoenix. There's gold here, perhaps lost mines, but also danger. Every year two or three people kill themselves exploring the mountains.
6. In the extreme south of Pinal County, 4 miles south of Oracle, are many old placer workings in the gulches of the Santa Catalina Mountains.
7. Mowry Wash, nine miles south of Patagonia in Santa Cruz County, was once a gold producer.
8. Yavapai County. There's gold near Prescott, all along the eastern foothills of the Bradshaw Mountains, in the creek beds. Some of the creeks (ask locally for detailed directions) are Humbug, French, Cow, Big Bug, Granite, Agua Fria, and Lynx.

Colorado

In addition to gold, Colorado has some fantastic scenery.
1. Chaffe County, all along the Arkansas River from Buena Vista (near the junction of US 24 and US 285) southeast to the Fremont County line. Also near Granite (17 miles northwest of Buena vista on US 24).
2. Douglas County. There are a number of gold mines within 50 miles of Denver. For example: Franktown, along Cherry Creek and Lemon Creek; Louviers, along Dry Creek; Parker, northwest 1.5

miles on route 83; and Elizabeth, along Gold Creek on route 86.

3. Gilpin County: Rollinsville along South Boulder Creek.

4. Gunnison County. Look in the north part of the county, especially along Washington Gulch.

5. Lake County: Along Box Creek; along the Arkansas River; around Leadville.

6. Montrose County: Along the San Miguel and Uncomphagre Rivers, and the creeks near Paradox.

South Dakota

Gold was discovered in South Dakota in 1874 by two miners attached to General Custer's force. The word got around and in 1876 there was a gold rush to the Black Hills, rivaling in activity the one in 1849 in California. Since then the state has been a constant producer; the Homestake Mine in Lawrence County is one of the nation's largest and richest gold producers. Practically all of South Dakota's gold production has been located in the relatively small area known as the Black Hills, in the southwestern corner of the state.

1. Lawrence County. Homestake Mine and others. Ask locally for more information.

2. Custer County along the banks of French Creek.

3. Try Rapid Creek near Placerville & Rockerville, both southwest of Rapid City.

4. Various Black Hills streams in the region due west of Rapid City, from Custer in the South to Lead and Deadwood in the north. Especially try Castle Creek, Crooked Gulch, Hoodoo Gulch and Chinese Hill, all near Mystic.

California Division of Mines and Geology

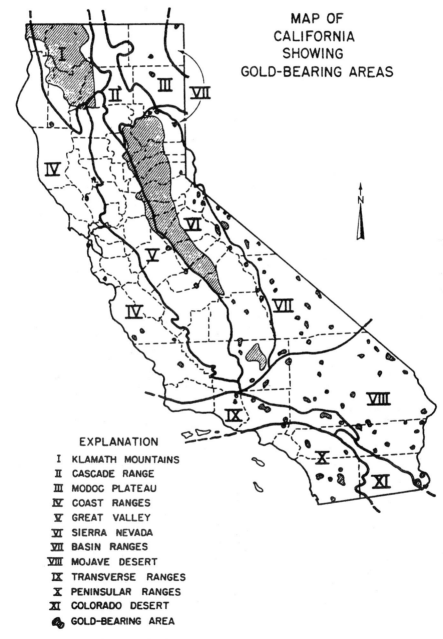

MAP OF
CALIFORNIA
SHOWING
GOLD-BEARING AREAS

EXPLANATION

I	KLAMATH MOUNTAINS
II	CASCADE RANGE
III	MODOC PLATEAU
IV	COAST RANGES
V	GREAT VALLEY
VI	SIERRA NEVADA
VII	BASIN RANGES
VIII	MOJAVE DESERT
IX	TRANSVERSE RANGES
X	PENINSULAR RANGES
XI	COLORADO DESERT
🜚	GOLD-BEARING AREA

Figure 1. California Gold Locations

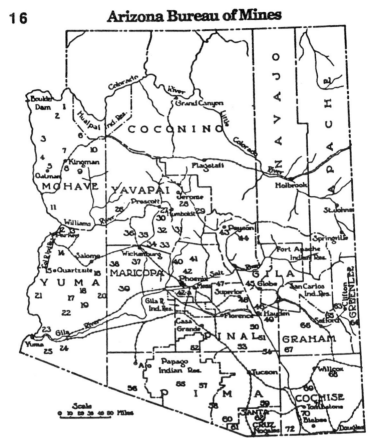

1 Lost Basin	25 La Fortuna	47 Goldfields
2 Gold Basin	26 Eureka	48 Superior (Pioneer)
3 Northern Black Mountains	27 Prescott, Groom Creek	49 Saddle Mountain
(Weaver, Pilgrim)	28 Cherry Creek	50 Cottonwood
4 Union Pass	29 Squaw Peak	51 Mammoth (Old Hat)
5 Oatman	30 Hassayampa, Walker, Big-	52 Casa Grande
6 Musie Mountain	bug, Turkey Creek	53 Owl Head
7 Cerbat Mountains	31 Black Canyon	54 Old Hat
(Wallapai)	32 Peck, Bradshaw, Pine	55 Quijotoa
8 McConnico	Grove, Tiger, Minnehaha	56 Puerto Blanco Mountains
9 Maynard	33 Humbug, Castle Creek	57 Comobabi
10 Cottonwood	34 Black Rock, White	58 Baboquivari
11 Chemehuevis	Picacho	59 Greaterville
12 Cienega	35 Weaver (Octave)	60 Arivaca
13 Planet	36 Martinez	61 Oro Blanco
14 Plomosa	37 Vulture	62 Wrightson
15 La Paz	38 Big Horn	
16 Ellsworth	39 Midway	63 Gold Gulch (Morenci)
17 Kofa	40 Agua Fria	64 Twin Peaks
18 Sheep Tanks	41 Cave Creek	65 Lone Star
19 Tank Mountains	42 Winifred	66 Clark
20 Gila Bend Mountains	42-a Salt River	67 Rattlesnake
21 Trigo Mountains	43 Payson (Green Valley)	68 Dos Cabezas
22 Castle Dome	44 Spring Creek	69 Golden Rule
23 Las Flores (Laguna)	45 Globe	70 Tombstone
24 La Posa	46 Banner or Dripping	71 Turquoise
	Springs	72 Huachuca

Figure 2. Former Gold Bearing Areas.

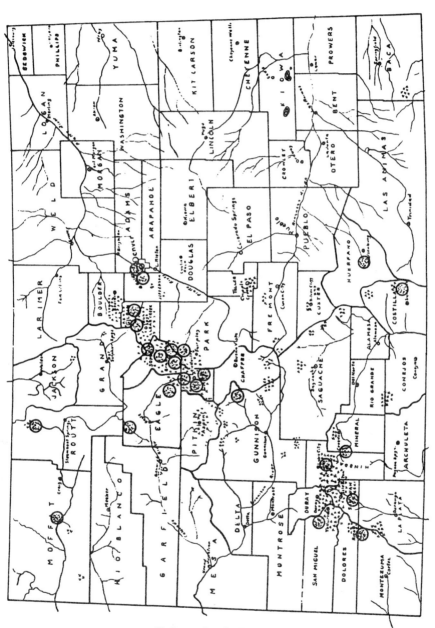

Figure 3. Colorado Gold Locations

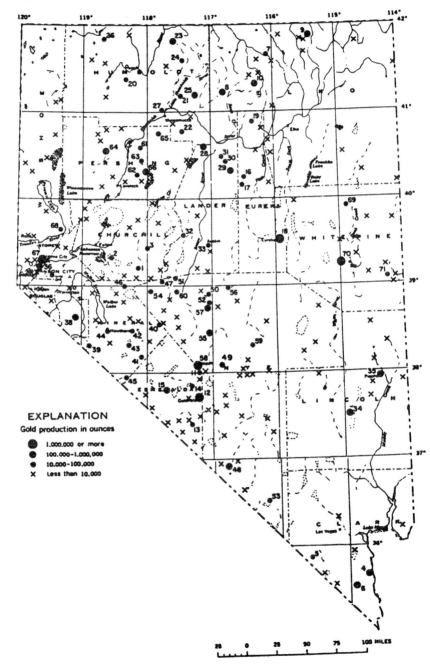

Figure 4. Nevada Gold Locations

Beach Gold

Not only is there gold in the mountains, there's gold along the beaches, too – sometimes. Beach prospecting is even more of a gamble than placer mining, but once in a while a prospector strikes it rich. It's been said that the beach at Nome, Alaska – only a few hundred yards wide – yielded two million dollars worth of gold before it was exhausted. Of course, that was in the old days.

Look for a place where a river flows down to the ocean from mountains known to have gold in them. This pretty much means the West Coast beaches, because the mountains must be close enough to the beach so that the gold is hustled on down to the beach before it settles out in the foothills.

The time to try is after a storm has raised the level of the stream so it floods down to the sea, bringing the gold with it. Look for your usual clues, especially the presence of black sand. But don't be misled by dirty or oily sand – the sand must be magnetic. If you find a place that fits all these requirements, get out your pan. And good luck!

CHAPTER 3

What to Look For

From a practical point of view, gold comes mixed with rocks. The rocks are pushed up as mountains by the heat and pressure deep inside the earth and then are worn down by the action of wind and rain. In the course of this erosion, the gold often becomes separated from the rocks, forming the rich deposits we're looking for. Then when the rain comes, the gold is washed down the creeks and the rivers, and lodges along the banks and behind the boulders of the streams. This was the source of the gold the college boys found in the Yuba River. This same process occurred in rivers that flowed thousands or millions of years ago in regions where changes in the geology have converted the once-fertile forests into deserts. This is where the deposits in the deserts come from, as well as the "pockets" of gold in the sides of hills.

This sort of mining is called "placer mining" and the deposits are called "placers". In "hard-rock" mining, the gold occurs in thin veins embedded in solid rock, usually quartz. Commercial miners look for these veins and then crush up the rocks in big mills and refine the resulting dirt with chemical means. Not very many mines are in operation these days because the remaining ones are difficult to work and the cost of extracting the gold makes the job unprofitable even at today's gold prices. These considerations do not worry the amateur placer miner, however, because his labor is vacation labor and he relies on Mother Nature to do the gold separation for him.

The kind of mountains that are likely to contain gold in any sizeable amount are either very old mountains or else very young mountains. For some reason, adolescent mountains do not seem to be as rich. Somewhat like adolescent people. In the United States this means that mountains of either Tertiary or Quaternary age are good prospects. To the geologist this means something, but the casual prospector can't tell a Tertiary from a hole in the ground. And asking the local inhabitants if their mountains are Quaternary usually brings only a blank stare. However, what it does mean is that the Rockies and Sierra Nevada — mountains west of the Mississippi River — are possible candidates. Some gold has also been found in the Appalachian Mountains, in Pennsylvania, Vermont, north Georgia, and Tennessee.

Given the right mountain range, the structure of a particular mountain is most important. Mountains, especially young mountains, are subject to earthquakes — a kind of geologic growing pain. Earthquakes produce "faults", which are the visible evidence of an earthquake. Faults are places where the tension of the growing mountain became so great that part of the ground pulled away from or slipped past the rest, leaving a long scar. The major fault lines, like the San Andreas Fault, are common in the Far West and are easy to spot once one knows what to look for. "Outcrops" are another evidence of past geologic activity. They are places where the basic underlying rock has been squirted up through the softer earth and then congealed there. The result is often a field or meadow of ordinary soil, pockmarked with monoliths and low ridges of solid rock. They are lines along which the normal slope of a hillside changes abruptly – a little like a closed zipper or a backbone of a fish. Locating faults is one way in which the local natives can help. For some reason they are usually quite proud of their earthquakes and are happy to point out the fault lines.

Figure 5. Fault Line

The fault lines and outcrops are important to the prospector because pressures allow minerals to be squirted into the rip in the earth. As a result, fault lines often point to gold. The normal processes of erosion may wipe away the visible evidences of the vein, but this same erosion may wash the gold from the face of the vein into a stream below. The prospector should look for gold below the fault, in a place where his judgement tells him that the rain might wash the gold. If he finds traces, he should work his way up to the fault and into it.

Prospecting for gold is a good deal like being a detective. The traces of gold in the pan are not the important thing, they're clues. Properly interpreted, they point to the source of the gold; an exposed vein or a rich pocket. And as in detective work, careful thinking and a knowledge of the "modus operandi" can take a load off the muscles. The basic principles are simple; The gold comes to the surface in the upward thrust of the mountains. It is washed toward the seas, and being very heavy, it settles to the bottom of the wash wherever the current is too weak to drag it along.

This characteristic settling action leads the wise prospector to pick spots along a stream where the current slows down or changes direction. The inside bend of a stream; the upstream side of a sandbar; the banks of a stream when the water is subsiding – these are all good prospects. The same applies to dry country, where the river has long since vanished but its outline remains. Don't forget that the desert is not dry all year around. Sometimes it rains there, and when it does the water rushes down the gullies in a real torrent. After such a rain, look along the dry banks as if the water were still there – try to visualize how the heavy gold might be carried. In particular you can now look in the middle of the former stream, where the biggest and heaviest bits of gold were carried and then settled out as the water subsided. Figure 6 illustrates some of these places.

Figure 6. Where to Look For Gold in a Streambed

1. On the inside edges of bends in the stream
2. In quiet pools
3. On the upstream side of obstructions
4. In the roots of river plants
5. Beneath whirlpools where two streams join

Here the gold will be deeper in the soil because its weight will cause it to settle, and it's had more time to sink into the ground.

Getting at the gold deposited in the bank at a bend in the former stream requires a little planning. The gold's inertia will drive it into the bank at this point, but over the course of time since the last flash flood the gold will continue to settle, while mud and dust will gradually build up the bank. This means that you should look somewhat back from the edge of the bend, and downward. Try digging down from the top of the present bank at the point where you guess the gold might have settled. Pan as you dig; at first you will get nothing, then presently you should begin to see to see color if your guesses are right. Keep going down into the main pocket. This process, by the way, is known as "drifting".

Figure 7. Look for Gold at the Barrier Level

This same settling means that there is rarely any significant amount of gold right on the surface. If the surface dirt is sand or gravel, the gold will settle through it over the course of time until it comes to an impervious layer. This barrier may be bed-rock, the basic rock layer stretching deep into the earth. Or it may be hard clay, or an old lava flow, or in fact anything that might impede the downward settling of the gold. If you find traces of gold near the surface, keep digging, and pan as you go down. The amount of color should improve as you near the barrier.

Another clue – gold is often found mixed with fine black or red sand. The black sand is a form of iron oxide. It is heavy and magnetic, and consequently quite easy to identify. The red sand, called ruby sand, is composed of tiny, crushed-up garnets. Although gold is often found with these sands, it is necessary to consider the way the black or red sand is spread around. If it is evenly distributed, the yield of gold per shovelful will be poor. On the other hand, if the red or black sand shows in sharply defined layers, the gold may also be concentrated in these strata.

Another clue is very obvious – so obvious that it is often misused. This is the presence of former mining operations. A lot of old mine shafts in a side of a mountain means that somebody once found gold there. Nature being what it is, there's a good chance there is still gold in the mountain. HOWEVER, you don't work the immediate vicinity of an old, tired mine shaft unless you're more interested in exercise than in gold. The old-timer who sunk the shaft was just as hungry as you, and perhaps a bit smarter at the game. He took the gold out until he decided that particular spot was no longer worth working. If he came to that conclusion, who are you to argue? Look over the situation; try to visualize the geology and apply the clues. Them move to the nearest spot where there's no shaft; maybe only fifty feet away. That new spot may be the very one the old-timer was planning to try next, except that a mountain lion ate him first.

DANGER! DANGER! Prowling around inside of old mine shafts may be injurious to your health. For instance, you may end up dead. Some of the hazards are obvious--falling down vertical mine shafts, which may be a hundred feet or more deep; rattlesnakes and spiders taking a little nap; falling rocks. The old timers used dynamite to blast the tunnels open; sometimes they left a few sticks of dynamite lying around when they left. Now, dynamite is a respectable, well-behaved explosive when properly used--but not when it's been lying around a long time. The nitroglycerine in the stuff will concentrate in one spot, making it liable to go off with even gentle handling.

Another hazard is still more subtle. It is the presence of "bad air" as it is known. Bad air has poisonous gases mixed with it; sometimes it is simply deficient in oxygen, having lain unchanged in the shaft for a hundred years. You can't smell it. As Bob Harrison, a geologist with the Bureau of Land Management says, "You don't feel it all. It just creeps up on you. You lie down or sit down and go to sleep." And never wake up.

The old miners used to carry a canary bird with them into the shaft. As long as the bird kept singing, the air was OK. If the singing stopped the miners moved out--pronto.

Not many weekend prospectors these days want to carry the family bird with them. A buddy works better; the person going into the mine has a rope tied around his middle, his buddy has the other end of the rope. When shouts and yanks on the rope get no response, the buddy, up at the entrance, hightails it to the nearest habitation for help. The buddy does not, repeat not go down into the mine after his friend. That would result in two dead people in the mine instead of one.

Yes, there are lost mines. In the wild and bloody days of the last century, a miner who walked into town with a sackful of gold from a new strike was a pretty bad insurance risk. There are many well-substantiated cases of prospectors who obviously hit a rich strike, never told where it was, and vanished mysteriously. Maybe they were dry-gulched by bandits; maybe they met with an accident. In any case they disappeared; the source of their wealth was never found. A later chapter of this book discusses lost mines, but for the present it is enough to say that a "lost" mine does not look at all like an ordinary abandoned mine. The lost mine is far from the beaten path, well concealed, usually overgrown with brush, and without much to see except a hole in the ground. Otherwise it wouldn't stay lost.

There is an extra dividend from exploring abandoned mines, however, if you're willing to look for something other than the yellow shiny stuff. The old timers were pretty careless, and all they were after was gold. Often one finds excellent gemstones in the rubble of an old mine's tailings. And sometimes the tailings contain precious space age minerals like beryllium, tantalum, platinum, etc.

To summarize the clues:

1. Look in young, growing mountains, especially if there is quartz present.

2. Look for fault lines and outcrops and search the gullies and ravines below them.

3. Along existing streams, look in places where the current slows down or changes direction.

4. In vanished stream beds, try to visualize where these places might have been.

5. Dig down until you get to a layer which would stop the downward settling of the gold.

6. Look for concentrated strata of black magnetic or ruby sand.

7. Use former mine locations as a guide, but don't scratch around in old mine tailings.

8. Don't get discouraged. You don't expect to hit the jackpot on the first try, do you?

CHAPTER 4

How to Get It Out

You have studied the gold locations maps and compared them with a highway map, and the condition of your budget. You have decided to follow the trail of the 49ers, up in the Mother Lode, in the foothills of the Sierra Nevada. What next?

We will assume you're accustomed to traveling and camping in that sort of country, and know what kind of clothes and equipment to take for a good vacation. What extras do you need for gold-hunting?

Basics

For basic equipment, you'll need:

A gold pan for every working member of the expedition.

A big shovel, and a very small shovel or oversized trowel to ladle the pay dirt into the pan.

A rock-hound hammer.

Some plastic medicine bottles to put the find in.

A plastic classifier with 1/2 inch diameter holes.

Two, five gallon plastic buckets.

A couple of wide-mouthed plastic containers.

An artist-type camel's hair brush, a small magnifying glass, a small magnet, and a compass.

One or more crevice-scrapers. More on these later.

A homemade suction pump. More on this later.

If you get really serious about this, you'll need in addition:

A pickaxe to properly churn up the ground.

A sluice or rocker, as big as you can carry in your car.

A little chamois bag, about the right size and shape to hold a golf ball.

Although the list looks formidable, the basic kit is very easy on the budget. It's a good idea to begin with it, and add the other items as you progress.

The gold pan is the foundation; it is to the prospector what a rifle is to a hunter, or a rod and reel to a fisherman. The pan looks like a dishpan that somebody sat on. The sides slope up at that peculiar angle for a very good reason – to let the unwanted gravel and water slurp over the edge, while the heavier gold stays at the bottom. It is kept nice and clean so the gold doesn't hide under some little fleck of rust. Remember, gold flakes are very small (that's what the magnifying glass is for).

Figure 8. Tools for Gold Panning

How To Pan Gold

You use the pan by squatting down beside a stream and dumping one of the little shovelfuls of dirt in the pan. Then you scoop up about a pint of water by deftly flicking the edge of the pan in the creek.

With this water, you puddle the dirt until it's soupy. Then you scoop up about a quart of water, and swirl the mixture 'round and 'round in the pan just like sloshing a martini around in a glass at a cocktail party. When you've got it going in a comfortable little whirlpool, you tilt the pan just a little and let some of the mixture slurp over the edge – away from you, unless you want a lap full of mud. You add some more water and continue until the ordinary sand and gravel and mud have gone over the edge, leaving only a stubborn streak of black or red sand. Then you drain off most of the water, and give the pan a deft little flick to spread the remaining sand evenly over the bottom of the pan. And then you examine it. If you see half a dozen little flecks of gold about the size of a pin head, you're doing all right. Dump the sand and the gold into the wide-mouthed plastic container for further more detailed separation at your leisure, and go ahead with your panning on a fresh load of dirt. Don't try to separate out the gold there in the field. There are more efficient methods available for use in camp, and anyway it's always a mistake to stop panning before you've exhausted your strike (or yourself). For some reason a strike never seems as rich when I go back to it after having done something else for a while.

One piece of advice – dig deep! Don't waste your time digging in the river itself, you'll do much better in the banks of the river. Gold settles, you know, and the deeper you go for your pan full of dirt, the better chance you'll have of finding gold. Keep going until you hit a barrier layer or bed rock.

If you want more information on the techniques of panning, a little book called *Gold Fever*, by Lois deLorenzo, has very detailed instructions and good sketches on how to pan. It's published by the Gem Guides Book Co., Baldwin Park, California.

Crevice-scrapers

Probably the best way to describe a crevice-scraper is to tell what it's used for. Most river bed rocks have cracks in them, called crevices. When the flood waters came booming down the river, the crevices got filled with dirt and sand and gold. Now that the water has gone down, it's your job to scoop out the contents of the crevices and pan the stuff for gold. The crevices can be anywhere from an eighth of an inch to a foot across. You can get inside the big one with your trowel, and inside the middle sized ones with a spoon you borrowed from your mess kit. But how do you scrape out the inside of a one eighth inch crack in a rock?

Everyone has their own solution to this problem. Some people file the end of a screwdriver down to a point, then bend the end down to make a little pointed hoe. Some people file down the edges of a teaspoon to make a little narrow scooper. Some people use a Vshaped wood chisel. Everybody uses a knife to scrape the sides of the crevice. You can also purchase a crevice tool at your local mining supply store. Crevice scraping can be a profitable way to spend your time. Put the gunk you scrape out into your pan, and pan it enough to get the mixture down to sand and pebbles. Toss out the pebbles and put the rest in your plastic container for future refinement.

Suction Pump

A suction pump can be used to get the stuff out of crevices, or for exploring the upstream side of rocks in a creek bed, or for other similar chores. Professional power suction pumps are available, of course, but they're expensive. It's easy to make a hand-operated suction pump yourself, as follows:

Get yourself a nice new grease gun or suction pump at an auto supply store or at one of the glorified "five and dime" stores. There are various types: one kind designed to be used as a grease gun, to squirt grease into various places on your car, and another meant to be used to suck up oil out of your car's transmission or motor. Either will work, but don't try to use the gooey old grease gun you already have out in the garage. Remember that gold loves to hide in oil and grease.

The gadget works like a giant hypodermic needle; you pull out the handle to suck up stuff from the crevice, and you push the handle to squirt it out into your plastic container. You will probably need to slip a piece of plastic tubing over the nozzle, to get down into the crevice. You may want several tubes, of varying length and diameter. At this point let your ingenuity be your guide. One suggestion--these things don't work very well on dry materials; if the crevice is completely dry, dump enough water into it to make the stuff gooey.

Final Steps

When you're ready to knock off panning and get the gold out of the remaining sand, here's how to do it. Get out your little treasure-bottle and put some water in it. The mixture of gold and sand is still in the wide-mouthed container. Give it a swirl to expose some of the gold flakes. Then take the camel's hair brush, moisten the tip, and scoop up one or more of the flakes of gold on the tip of the brush. Carry them gently, still on the tip of the brush, to the little bottle and dunk the brush in it. The gold will fall off the brush into the water. Although this sounds complicated, the process is really quite simple once you get the hang of it.

If the sand is magnetic, as it usually is, you can make a preliminary sorting in the wide plastic container by moving the little magnet around right under the container. Always use the magnet three times on dry black sands, as small flakes tend to be picked up in each pass. The magnetic sand will follow the magnet, but the gold will not, so you can get most of the sand off in a little pile by itself. This makes it easier to get at the gold with the brush.

The brush method is not the only way of picking gold out of sand, but for people with big fingers it is often the most convenient. Ladies sometimes use their long fingernails to scoop up the gold; or tweezers can be used, or a pair of pointed chopsticks like the old Chinese miners used to do. But I've found the brush to be the fastest and surest.

Sometimes the pan will show very tiny flakes of gold, so tiny they're hard to see. This is called flour gold, and the flakes are so small because they've been trapped in a whirlpool or under some roots, and scrubbed over and over by the sand and gravel washing down the stream. Move farther up the hill, in the direction you think the gold came from. When you do so, the panning will be more erratic; some pans may show nothing, but others will show larger flakes.

Gold is often embedded in rock such as quartz. The gold may be mixed in the rock in such a way that its presence is not apparent, or traces of gold may be seen. In the latter case, the rock often has a "wormy" look. Crush the rock by breaking it up and then pulverizing it in a mortar and pestle. Then pan the resulting sand in the usual way. Incidentally, pieces of quartz are sometimes found with veins of gold in them, but without the wormy look. Don't smash these up – they can be very beautiful when cut and polished. In the old days, good milky quartz with embedded gold brought higher prices than normal gold.

Another way in which the gold hides from you is to get itself coated with a dark rusty exterior rock. This sometimes happens with small nuggets from the southern part of the Mojave Desert. If you have any suspicion that the dirty, rusty pebbles you're holding may be gold (perhaps because it's abnormally heavy) try roasting it. Put it in an old frying pan, or a G.I. can lid, or on an old shovel, and hold it over the campfire until it gets red hot. Keep it that way for about twenty minutes or so, or until it stops smoking. This will drive off the rusty stuff and you can see what you've got. Stir the roast from time to time, but with a long stick because rocks sometimes explode when heated. Also, since gold and natural mercury are often found together in nature, be sure you are upwind when heating the ores as mercury vapors can be fatal.

Figure 9. Roasting Ore

Amalgams

Now, if you're really going at this business seriously, you will not long be content to pick flakes of gold out of sand when you could be spending the precious hours shoveling dirt. If you've graduated to this sort of production-line mining (and it probably means you're using a sluice or rocker instead of a pan), you'll have to amalgamate the gold and then recover it.

First a warning about working with mercury. Mercury in this day and age along with the problems involved is best left to large scale miners. It only takes .01 mg in the human body to kill a person. Once in the body mercury never comes back out. There are also undocumented reports of stiff fines for the use of mercury in state and federal recreation areas as it is a tremendous hazard when lost into the environment. It is best to save the concentrates until you have several and then take them to someone knowledgeable to do it.

If you decide to amalgamate your own gold this is how you do it.

Put a little blob of mercury, about the size of a pea, in a wide-mouthed container, or whatever you're storing your concentrated gold-sand mixture in. Don't put the mercury in your prospecting pan, or you'll end up with a thin coating of mercury all over the bottom of the pan, which you'll have to scrape off. Roll the mercury around in the bottom of the container, which should be reasonably free from water. The gold will dissolve in the mercury, forming what's called an "amalgam". The sand will not dissolve. When all the little flakes of gold have been taken up by the mercury, it will have a yellowish and sort of bulgy and floppy look.

Now you have to get the gold out of the mercury. The old-timers used to squeeze it out, using a little chamois bag. Put the blob of mercury in the bag, tie up the top and squeeze down from the top. A good way is to twist the bag tightly. The mercury will ooze through the pores of the chamois bag, leaving a button which is mostly gold, but still partly mercury. The remaining mercury can be driven off by heating the button slowly in a pan over a low fire for half an hour or so.

*** WARNING: MERCURY AND ITS FUMES ARE POISONOUS. ALL HEATING OPERATIONS SHOULD BE DONE IN THE OPEN, WITH CARE NOT TO INHALE THE VAPOR. DON'T USE THE PAN FOR COOKING, AFTER IT HAS BEEN USED TO VAPORIZE MERCURY.

The second way the old-timers did it was to use a white potato, which they cut in half lengthwise. They scooped out a little hollow in one half and put the blob of mercury amalgam in it. Then they baked it an hour or so over a slow bed of coals. The potato absorbs the mercury, leaving a little button of gold.

*** WARNING. DO NOT EAT THE POTATO, NOR LEAVE IT AROUND WHERE CHILDREN OR DOGS CAN GET IT. BURY OR BURN IT.

Sluices

If you are interested in some method that gets the gold out of the ground faster than panning, a sluice or rocker is the answer. Panning yields most of the gold in a shovel full, but it is slow. Sluicing works fastest of all but looses the smaller flakes and needs a lot of water. Rockers are intermediate—faster than a pan, but less water wasteful than a sluice. Use a pan to explore; then if you find a good location, set up a sluice or rocker and start throwing dirt.

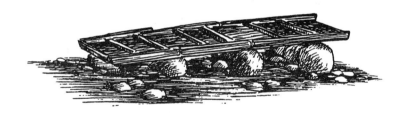

Figure 10. Sluice

A sluice works on the same principle as the bottom of a river bed. A stream of water flows over a set of riffles, or obstructions, sticking up from the bottom of the sluice. You shovel dirt on the head end (where the water goes in). The water washes it down the sluice. Because gold is heavier than most other ingredients in the sand and rock, gold settles behind the riffles. A

screen and some cross bars (the "growlers") are set up
at the head end to catch the rocks that may be mixed in
with the dirt, and keep them from banging along the
sluice and knocking out the riffles.

A sluice is easy to use. You set the head a little
higher than the tail. (Pick the best angle by trial and
error – try 10 degrees to start) and you dig a trench so
the tailings coming out the bottom won't pile up and
block the flow of water. You set the head near the
stream so you can divert some of the stream water into
the sluice, and you shovel dirt into the head end,
through the growler. Water washes the dirt away; the
gold gets stuck against the riffles. You shovel and
sluice until a) you're exhausted, or b) the sluice is full
of gold. Then you stop the water and get the gold out of
the sluice, picking it out or amalgamating it with
mercury, or scraping the sand out from behind the
riffles and panning it.

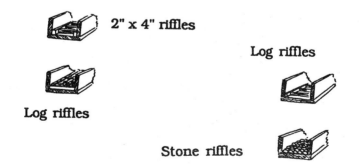

2" x 4" riffles

Log riffles

Log riffles

Stone riffles

Figure 11. Some Arrangements of Riffles

Commonly used today are 3-4 1/2 foot aluminum models, although they require cleaning two to three times a day, they are easier and more manageable to handle than the wooden homemade ones. A sluice is not hard to build. The nature of the riffles is not important; the length of the sluice is. The longer the better, with eight or ten feet being the practical minimum. Cars being what they are, sluices are often built in shorter sections for transportation, and then bolted together on the location. Don't worry about making the sluice absolutely water tight, but be sure the joints where the sections overlap are bolted together in such a way that the water flows over the joints without obstruction. Otherwise you'll have an unwanted riffle with a leak in it, and the gold will dribble back on the ground. The book mentioned earlier by Lois deLorenzo (*Gold Fever*) has a good description of how to build a sluice.

If you are planning to work in a place where water is not so plentiful, a modification of a sluice called a dip box may work for you.

The dip box is a short sluice. You can make one by nailing sides of 1" x 6" wood six feet long onto the sides of a six foot 1" x 12" floor. Nail another 1" x 6" board along one end to form a back. Cover the bottom of this trough with carpet or burlap to catch the gold. You'll need riffles to slow the runoff and allow the gold to settle. An easy way to make riffles is to set a sheet of wire mesh over the bottom five feet of your carpeted trough. Give the dip box a steep slope by setting it up on trestles like uneven saw horses.

Figure 12. A Dip Box is a short sluice that can be used where water is scarce.

To use the dip box, just shovel dirt into the top foot of the trough and wash it over the riffles with buckets of water. Don't run the water too violently or you'll wash everything away. You'll find that you have to remove the larger rocks from the trough by hand to use the box successfully.

Rockers

A rocker is sort of a folded-up sluice with refinements. It is easier to move around than the sluice, and can process more dirt than a pan. An average miner finds it hard to process more than half a cubic yard of dirt a day, using a pan. However, two men using a rocker can run from two to five cubic yards of dirt through in a day.

Rockers are built in three distinct parts, a body or sluice box, a screen, and an apron. The floor of the body holds the riffles in which the gold is caught. The screen catches the coarser materials and is a place where clay can be broken up to remove all small particles of gold. The apron is used to carry all materials to the head of the rocker. A rocker is like a sluice which is rocked back and forth sideways to toss the gravel around more and give the gold a better chance to settle. It also has a piece of nappy carpet, known as the "apron" along which the gravel flows. The carpet nap catches the flakes of gold and holds them. The 49ers built their own, and each miner built his differently. The main principles are that the apron is set at a fairly steep angle, like 30 or 45 degrees. The final sluice section is optional.

Figure 13 shows a box, open at one end and at the top. It has a removable false bottom (not shown), with some riffles in it to act as a sluice for the gravel that gets past the apron. The main part of this contraption, however, is the apron, which is a piece of ordinary carpet laid loosely in the box from the upper left of the figure to the lower right. That is, the apron goes from the top of the open end to near the bottom of the closed end.

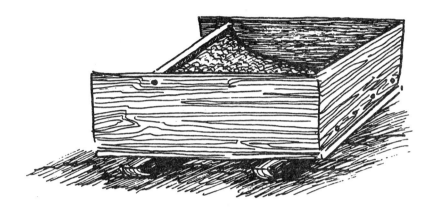

Figure 13. Rocker

The apron is set in on a couple of 2 x 4's going across the box at the top of the open end and the bottom of the closed end. It is set in so that when you dump dirt and water in at the top end, the nap of the carpet will catch the gold flakes. To make the apron removable for cleaning, fasten it to a couple of narrow boards, and fasten these to the 2 x 4's at the corners of the box.

The last step is to put some rockers on the bottom of the monster, so it can be rocked back and forth sideways. This helps to catch the gold in the carpet nap. A couple of boards sawed with a curve in them make fine rockers — or you can just put some pegs in the bottom of the box and let it go "bump-bump" when you rock it. The gold won't mind.

In operation, one man shovels dirt onto the screen and sees to the supply of water, either by diverting a stream or by bucket supply. The other man rocks the machine to and fro with a jerky motion. The water washes the dirt down through the holes in the screen on to the apron and on down the apron to the sluice section and then out the back door. When the rocker gets loaded up with sand (and gold, hopefully), the rocker must be cleaned out. The apron is removed and is washed carefully in a tub of water, where the gold settles to the bottom. The sand behind the riffles is carefully scraped and brushed out. Then the sand and gravel from both the apron and the sluice section is panned to get the gold, or mercury is used.

A mini-model can be built or purchased and is recommended for single person operations.

Although both the sluice and rocker require panning or amalgamation as the end process to get the gold, panning the concentrate from one of them is much more satisfying than panning raw dirt. In fact, this is probably the closest the modern prospector will ever get to the rich pay dirt of the gold rush days, where according to the *Encyclopedia Britannica,* a simple panful of river gravel sometimes yielded as much as 15 ounces of gold! Ah, for the old days!

Stream Prospecting

Don't forget about hunting for gold along the bottom of streams. The streams should be those running down from the mountains likely to have gold; the best time to prospect the bottom is right after a big spring rain, when the rushing water has been tearing the soil and creates flood layers.

A stream acts just like a natural sluice; the rocks and other obstructions on the bottom catch and hold the gold on the upstream side. A simple way to go after the gold is simply to dive with a snorkel or face mask, carrying a small bucket and a trowel. Dig out as much gravel as you can from the upstream side of big rocks before your air gives out, then swim back to shore and dump the bucket in some sort of container for later panning.

Or you can rig some sort of homemade dredge to suck the dirt out from under the rocks, or use a home-made giant syringe made out of a grease gun, as described in *Gold Fever*. Your ingenuity is your only limit.

CHAPTER 5

Testing — Testing

"Fool's gold" is the bane of the beginning prospector's life. He can see lots of little flakes in the bottom of his pan that glitter enticingly, especially when the sun hits them. In fact, even the dirt on his shovel is speckled with the beautiful stuff – even before he starts to pan it. Tough luck! He's probably hit a bonanza of fool's gold, worth exactly nothing.

Fool's gold is iron pyrite, or iron-copper pyrite, or a form of yellowish brown mica. At first glance it looks very much like gold, particularly in a strong light, or when viewed under water. However, it's not hard to tell it from the real thing with a few tests.

In the first place, gold retains its yellow sheen even in shadow, whereas fool's gold does not. This is the quickest and simplest test and it means that the pan should be examined in the shade, not in sunlight.

In the second place, gold is soft and malleable. You can cut it with a knife without splintering it, and you can poke a dent in it with a pin. Fool's gold is flaky and hard, like mica. If you try to cut it, it will splinter or flake.

In the third place, it's possible to see the flaky structure of fool's gold with a good magnifying glass. Gold seldom has any crystalline structures; it looks like a little flat blob instead of a flake, although for convenience we speak of gold "flakes".

There are a number of chemical tests. The simplest is to put a drop of nitric acid or hydrochloric (muriatic) acid on the sample. Gold will not be affected, but fool's gold will react with a slight foaming action and will finally dissolve. Muriatic acid can be obtained from drug stores, or from swimming pool supply stores.

SPECIFIC GRAVITY OF COMMON METALS

Mica	2.3
Feldspar	2.5
Quartz	2.7
Calcite	2.7
Hornblende	3.2
Garnet	3.5
Topaz	3.5
Corundum	4.0
Magnetite	5.2
Silver	7.5
Lead	11.3
Gold	19.2

Gold is one of the heaviest elements known. When pure, it has a density (specific gravity) of about 19, which means that it weighs about 19 times as much as an equivalent amount of water. If the gold is mixed or alloyed with other minerals, the average density will be less than 19. Consequently, specific gravity measurements can give a rough estimate of the purity of the gold. The average specific gravity of some common minerals is given in the table.

The first five are the common constituents of the sand and rocks found associated with gold. Their low specific gravity is the reason they wash away in panning. Magnetite is typical of the magnetic black sand left at the bottom of a pan mixed with gold. Topaz and corundum, along with garnet, form the "ruby" sand sometimes found at the bottom of a pan. In big enough pieces they are gem minerals; corundum appears as ruby or sapphire, depending on the impurities in the stone. Topaz is an amber or yellowish gem, and garnets are red, as is ruby.

In some regions, these minerals are found in the same localities as gold. Keep your eye out for any abnormally heavy rock. It might be one of these three, or it might be quartz with gold or silver inclusions in it.

Because of the usefulness of the specific gravity test, here's a brief outline of how to do it. Specific gravity is found by dividing the weight of the sample by the volume of the sample. The weight is easily found with the aid of a balance or scales. The volume is found by using a trick discovered by an old Greek named Archimedes.

To make specific gravity measurements, you need a balance capable of weighing in fractions of a gram. (A gram is a very small weight, about 1/28 of an ordinary ounce. If you have a chunk of gold big enough to weigh on ordinary scales, you don't need this book. Sell the chunk and retire.) The balance must also be arranged so you can weigh the sample both in the normal way and when the sample is dunked in a little glass of water. (The glass of water must be resting on the table, not on the pan of the balance.)

The reason for this is that you need to find out how much lighter the sample seems to be when it is immersed in water. Everybody has seen and felt the bouyant effect of water, when they're swimming in the ocean or in their backyard pool. Sometimes the bouyancy is so great we can actually float in the water. The degree to which the water bouys us up is determined by our specific gravity--our weight in relation to our size.

To determine the specific gravity of your little gold nugget (or bag of gold dust, or whatever) you need to know its normal weight and its apparent loss of weight when it's immersed in water. The first of these is easy--buy a good scale which reads in grams, not ounces, and weigh the nugget. (Using grams simplifies the arithmetic.)

To get the apparent loss of weight in water, we must weigh the little nugget again. But this time we place a container of water in front of the scales in some spot where it would be convenient to dangle the nugget in the water. You may have to tie one end of a piece of thread around the nugget, and hang the other end of the thread from the balance beam of the scales. Or maybe you have some other arrangement of the experiment. The only important point is to be sure the scales are weighing the nugget, the whole nugget, and nothing but the nugget.

Figure 14. Finding Approximate Specific Gravity

Well, now you have the weight of the nugget in air and the weight of the nugget when immersed in water. The difference is the loss of weight in water. Now you put these two numbers into a simple little equation as follows:

$$-\frac{\text{Sample weight in air}}{\text{Loss of weight in water}} = \text{Specific gravity}$$

For example, suppose your little nugget weighed 34 grams in air and only 32 grams in water. The loss of weight is 2 grams. Putting these numbers into the equation we get:

$$\frac{34}{2} = 17$$

Your nugget has a specific gravity of 17. It is gold, but not pure gold, which would have a specific gravity of 19.2.

There are various other tests, some more accurate, some less, some simpler, some more complicated. One of the simplest , and least accurate, is the color scratch test. It is based on the fact that the color of gold varies, depending on the purity of the gold. Many rock shops sell what they call "scratch kits", consisting of a "master color chart" showing the colors of gold for various levels of purity. You make a scratch with your gold sample on a piece of sandpaper and compare the color of your scratch with those on the master chart. It certainly is a simple test, but I wouldn't rely on it very much.

There are many other tests that can be made on unknown minerals to aide in identifying them. Collecting rocks and minerals is a fascinating hobby itself. For more information, the reader is referred to a very handy book, *Rocks and Minerals*, by Zim and Shaffer, published as a pocket guide by Golden Press, New York. It is full of excellent color pictures and descriptions of most of the minerals you are likely to find. Sample kits of rocks and minerals can be bought from most rock shops and hobby shops, and are a good way to get an elementary knowledge of mineralogy.

CHAPTER 6

Mining In the Desert

I am convinced there is just as much gold waiting to be taken out of the desert mountains as out of the Mother Lode region. The desert mines just never got the publicity. For one thing mining in the desert is a great deal harder. For another there just haven't been as many people prospecting the desert. Nevertheless, a lot of gold was taken out during the last of the 19th century, and the first part of this one. The mines around Panamint, the ones around Randsburg, and those near Mojave and Twenty-nine Palms produced among them well over $50,000,000 in gold.

Desert prospecting requires some different tricks from stream prospecting. For that matter, travel and camping are different in the desert. Here are some basic rules for the newcomer to desert life. They apply particularly in the summer, but they're good to keep in mind all year 'round.

1. In summer, any part of the desert off the main paved road is dangerous. It is very easy to get stuck in the sand, and the side roads are not much travelled in summer. Before you go out, leave word with a friend about where you're heading and when you expect to be home.

2. If you get stuck or the car is disabled, stay with your car. It will afford some shade and makes a good landmark for searching parties.

3. Carry at least a gallon of water per person per day. A man won't starve to death for a long time, but he can die of thirst very quickly.

4. Wear a hat and light clothing that covers all your body. That sun is hot enough to fry an egg, and you, too. Carry sunscreen to prevent burning.

5. Unless you know what you're doing, don't try any fancy tricks of recovering water from moist sand. It can be done, but you can also exhaust yourself trying. Wait by the car for help.

6. While you're waiting read *Desert Survival* by Primer Publications, Phoenix.

Now that I've said all this nasty things about my old friend, let me repeat that I love the desert – in fall, winter and spring. In summer, I cross it as quickly as possible in an air-conditioned car. During the cool months it is truly delightful – cool at night and warm by day, interesting, silent, smog free, and people free.

You can see a long way in the desert, and you can spot a likely-looking place to prospect from quite a distance. The black or red hills are the good bets. And if you see the little holes or abandoned shacks of former miners dotting the flanks of the hills, try your luck near them.

Desert mining is complicated by the lack of water. The old timers used the wind to separate the gold from the sand, by tossing the gravel up in the air and catching it in a blanket like an old time farmer winnowing the wheat. The gold, being heavier, fell in the blanket, while the wind blew away the sand. They also built a number of odd contraptions like rockers with bellows on them, to blow air over the apron. Most of these weird machines provided more exercise than gold, mainly because even desert sand is a little moist, once you get below the surface. The gold stuck to the damp grains of sand, making the separation very poor.

The modern prospector, by carrying extra water in his car or truck, can use the same equipment as for stream panning, with a few modifications. Panning or rocking are the best ways; sluicing just takes too much water. The modifications consist of arranging some way to catch the water after it has been used, instead of throwing it away. For example, in panning, a big old-fashioned wash tub and a couple of five-gallon cans of water will take care of a weekend's panning. You simply squat in front of the tub, holding the pan down in it, and pan in the usual way, taking care not to bark your knuckles on the sides of the tub. With the rocker, you caulk it up to make it more watertight than usual, and you put the tub at the tail end, where it will catch the gravel and water. In both these operations the tub gets full of mud and sand pretty soon. Then you pour the water back in the cans, dump the mud out of the tub, and start over. Bit of a nuisance, but it works.

One of the bonuses of mining in the desert is the abundance of interesting rocks. Many of these are semi-precious gem quality, ranging all the way from unusual stones like the silver onyx of the Calico Mountains to the gem quality aquamarine and tourmaline found near Mt. Palomar and the turquoise from the mines near Baker, CA. There is an abundance of agate and jasper all over the desert, and some excellent petrified palm root in the hills southwest of Mojave. Nodules, which look like petrified hen's eggs, can be found less than an hour's drive from Los Angeles, in the region near Saugus and Moorpark. These nodules, when sawed in two and polished, make interesting and beautiful paper weights.

What to do with the rocks? There are dozens of lapidary shops in every city which will be glad to saw them up and polish them, or mount them as jewelry, for a small fee. The writer, who became interested in lapidary work about the same time as the gold bug hit him, has many cuff links, tie clasps, and bola slides made of onyx, agate, and jasper found on trips to the desert. Some of them show "pictures" formed by impurities which seeped into the rock when it was being formed. The precious stones, like tourmaline, aquamarine, or turquoise, are more rare, and make up into beautiful jewelry.

If the weekend prospector happens to find a mine of one of these latter stones, he may well make more money from it than from the gold he might find. The stones appear in "free" form (loose on the ground) so little processing is needed to get them out, and good quality rough stones bring a nice price. It would be worthwhile for the prospector to visit a rock shop in a desert town to get an idea of what these stones look like. Most rock shops now carry an assortment of books on rocks and minerals as well as gemstone identification, some of which pertain to the local areas.

CHAPTER 7

So You Want To Do It Electronically?

In this day of atomic energy and TV pictures of the surface of Mars, it may seem an anachronism to go out hunting gold with a shovel and battered old pan. There ought to be some sort of electronic marvel, preferably with a built-in computer, that would let the prospector relax under a tree while the machine scanned the terrain like a bird-dog looking for quail.

I'm glad to say there isn't. I'm getting a little tired of the marvels of science, and when I go out prospecting, I want to dig and sweat and let my beard grow and forget about civilization.

There are, however, a few gadgets along the electronic line that can be useful. None of them eliminates the need for human judgement – thank goodness – and none of them keep the prospector from being tired at the end of a day's work. But they do provide clues.

The simplest of these, and maybe the most useful, is a metal detector, marketed under various trade names you can buy at prices from $39.95 to $700.00 and up. I often take one of the lower priced varieties with me when I go out into gold country, partly to help find good clues to gold, and partly to hunt for lost treasures in old ghost towns. A later chapter covers the ghost towns, and the present discussion will be confined to the use of the metal detectors in gold hunting.

They look like army mine locators-- a pie tin on the end of a broomstick. They're usually prettied up with nice cases, streamlined handles and fancy indicators. You can buy them at all treasure hunting shops and most of the retail electronics shops like Radio Shack. Sears Roebuck sells eight or ten different models, ranging in price from about $250 to about $650.

Most of the early metal detectors operated on one of two different principles: the beat frequency oscillator principle (BFO) or the induction balance (IB). The beat frequency oscillators (BFO) were very easy to make and inexpensive--an experienced ham radio or stereo fan could throw one together using parts from his scrap box. The resulting device would usually be moderately sensitive but have very poor discrimination. You couldn't tell a penny from a can of beans, which made for a lot of extra work for the prospector.

Virtually all modern detectors operate on the induction balance, or IB principle. The search head contains two or more coils which are carefully arranged so as to be electronically balanced. This is often called the transmit-receive, or TR system. A metal object, brought near the loop, causes an imbalance that signals a find. Early models, called metal/mineral TR detectors, respond to changes in the amplitude of the receiver signal. They, like the BFO, could tell the difference between ferrous and nonferrous objects, but lost depth in mineralized ground, where most gold is found.

Some IB instruments can be adjusted to neutralize the effects of ground mineralization, allowing them to detect deeply buried metal objects which could not be found with the BFO or metal/mineral TR. These detectors operate in the 3 to 30 khz very low frequency, or VLF, range. Most have a discrimination mode which can reject certain trash items, such as nails, bottle caps, foil, pull tabs, etc. The discrimination mode is sometimes called the TR.

VLF detectors are available with a wide range of features, among them being automatic ground balance, audio and visual target identification, discrimination while ground balancing, target depth meters, etc. Some models contain an on-board computer and can be programmed to accept only those objects the operator is interested in. As a general rule, those instruments which can operate in the 4 khz range are capable of excellent depth in highly mineralized ground and are highly sensitive to silver and copper, but less sensitive to gold. Detectors which are designed specifically for electronic prospecting usually operate in the 20 khz to 50 khz range. They are capable of finding very tiny nuggets; however, they may not be able to achieve great depths in heavily mineralized ground. Uncontrolled discrimination is not desirable in a gold hunting detector because it would tend to reject the smaller nuggets.

The instruments are used by waving the search coil back and forth over the ground. The ground balance is adjusted to cancel the effects of mineralization so that, when the coil passes over a buried metal object or hot rock, an audio tone is heard. (Hot rocks are highly mineralized rocks which can give a positive response.) Electronic prospecting, or nugget hunting should be done in an all-metal mode so that nothing is missed. Tiny nuggets can be rejected at low levels of discrimination. Along with gold, the detector will find nails, tin cans, pull tabs, bullet fragments, scrap iron, etc. It will also detect any coins and jewelry which some former visitor might have dropped.

Another use for a metal detector in gold hunting is in locating streaks of black sand. This sand usually is slightly metallic, and will make the detector react. The detector is especially useful in places where the wind or water tend to strew white sand over the black, so the latter is hard to see.

Two other cheap electronic gadgets are worth carrying with you. One is a transistor radio. It is not entirely for entertainment. On the desert, storms come up amazingly suddenly, and dump a tremendous amount of water in a very short time. So it's a good idea to have the little transistor radio turned on while you work. When the snaps and crackles drown out the programs, it's time to move before you get drowned out.

The other gadget is a cheap, short-range walkie-talkie or C.B. transceiver. If a whole family is prospecting all over a mountain side, it is very frustrating for one member to hit a good strike and not be able to contact the rest of the family. These gadgets are also a safety measure, in case one member has an accident.

CHAPTER 8

What to Do If You Strike It Rich

Assuming you've found a likely looking area and you want to start prospecting – how do you know you're not trespassing on someone's property? If the area is fenced or posted, stay out. Secondly, don't prospect in the National Parks or National Monuments. Thirdly, if the place looks like uninhabited wilderness, it probably is. If you wouldn't worry about fishing there or putting up a tent for an overnight stay, go ahead and dig.

A rich find for a weekend prospector usually consists of a half-ounce of dust or a few nuggets. However, once in a while, he finds a really good deposit – too rich to exhaust in a single visit. If the location is on public land, it is possible for the finder to establish rights to the location. The laws on this vary from state to state, and are very complicated. Check with a lawyer – if the find is rich enough, it's worth a lawyer's fee.

Here are a few general rules and guides about filing a claim. The local and federal governments both get into the act. The local point is the county recorder where the claim is located. The federal government is represented by the Bureau of Land Management of the Department of the Interior. It issues a pamphlet which tells how to establish a claim, and it also has other publications especially one entitled "Camping on the Public Land." The Bureau of Land Management has offices in the various states, as follows:

Arizona: Federal Building, Room 3022,
   Phoenix, AZ 85067

California: Federal Building, Room 4017,
   Sacramento, CA 95816
                    or
1414 University Ave., Riverside,CA

Colorado: Federal Building, Room 14023,
   Denver, Colo. 80105

Nevada: Federal Building, Room 3008, Reno,
   Nevada 89505

Oregon: 729 N.E. Oregon St., Portland,
   Oregon 97026

For other states, write to the Bureau of Land Management, Department of the Interior, at the state capitol. The post office will get your letter to the right place. Another good source available in book and prospecting shops is the book *Stake Your Claim.*

In general, rights to a mineral deposit in the public lands can be established by prospecting and "locating" a find. In some cases the land can become the property of the finder, but only for the purpose of mining. In any case, he has exclusive rights to the mineral deposits on his find. "Locating" here means staking the corners of the claim area, posting a notice, and complying with the state laws about recording the location with the county recorder's office. The prospector is permitted to claim 20 acres, laid out along the lines of the public survey (i.e. along the section and township lines in the official map of the region), if he should be claiming a placer deposit. The rules for claiming a vein are a little different. The corners of the claim should be staked with little monuments of stones, and the notice should be fastened to a post with an empty can or wooden box over it to protect if from the elements. The notice should contain as a minimum the date, the county, state, township, and section where the claim is located, the number of acres claimed, and the name of the claimer. However, anyone seriously considering filing a claim should consult the pamphlet issued by the Bureau of Land Management in the state in which filing the claim and also check with the county recorder prior to posting a notice.

In any case, a prospector who finds a rich deposit should make sure he can locate it exactly on an official map. Detailed maps are available at ranger stations in the National Forests, or from the Superintendent of Documents, Government Printing Office, Washington, D.C., or from private concerns. Once you have a map of your area, locating exactly where you are requires a compass, a mileage indicator, and some common sense.

Locate on the map the nearest pinpoint identification spot, like the junction of two streams, a road intersection, or a sharp bend in the trail. Then, using the compass and the mileage indicator on the car, if you're driving, determine how far the location is and in what direction, from the identification point. If you're on foot, you will have to estimate the distance, remembering that a good stride by a man is about one yard. The maps are usually divided off in squares of one-half or one mile on a side, so you should be able to spot the location to a fraction of a mile.

One word of advice. A treasure-filled location is often not very big – if you miss it by twenty-five feet, you've lost it. You do not have to describe the location that precisely in filing a claim, since you're allowed to claim a good-sized chunk of land, but you may discover that the following weekend, when you come back to work the deposit some more, you simply can't find the exact place. This, incidentally, is one reason so many mines get lost. A small hole in the ground in rough country is very hard to find if it's not marked.

So you have to mark the location with a pile of stones or something that won't blow away. And as you're working your way back to the nearest map checkpoint, you mark your trail the same way. Otherwise you'll contribute one more little legend to the lost mines of the Golden West.

The gold you bring home with you deserves a better fate than a little medicine bottle. You can sell it, if you want to. The U.S. Mint, in Washington, will buy gold in quantities over one ounce, troy weight. (Troy weight is a system used in weighing precious metals. One troy ounce equals 1.09714 avoirdupois ounces, and one troy pound equals 0.8228 avoirdupois pounds. There are 12 ounces in a troy pound instead of the 16 ounces in the ordinary avoirdupois system. Incidentally, you can see why we want to go from this weird system of avoirdupois and troy pounds and ounces to the metric system, where everything is a simple multiple of ten.)

You can also sell your gold to accredited dealers, who will buy it at about half to 2/3 of the price quoted on the international market (at this writing, about $385.00). Check with a "manufacturing jeweler" in your town.

Or you can have your gold made up into jewelry or souvenirs. Gold dust or nuggets embedded in a plastic block make very fascinating conversation pieces as paperweights. Or you can have the gold embedded in a plastic teardrop to be worn as a pendant, or mounted under a little plastic or quartz dome and set in a ring or a tie tack. Gold embedded in a clear plastic pen-holder desk set is also very effective. A very fascinating coffee table can be made by taking thin strips of the unusual rocks you find on your trips, plus the gold and snapshots of the great discovery and embedding the whole works in plastic as a coffee table top. The possibilities are endless, and you can do this embedding bit yourself – it's fun. But don't just let the gold sit in your dresser drawer. Gold is much too fascinating for that.

CHAPTER 9

Ghost Towns and Lost Mines

Ghost towns of the West range all the way from completely abandoned villages to authentic reproductions and restorations – complete with inhabitants dressed in mining clothes and Mother Hubbards, and with a shoot-out every hour on the hour. The restored towns are easier to get to, and they're a lot of fun, except that I always have the slight feeling of being in Disneyland. One of the better ones is Calico, near Barstow, CA , which has been restored authentically and lovingly by the Knott family of Knott's Berry Farm. Another is Virginia City, near Reno, Nevada. Anyone who has read Mark Twain's *Roughing It* has a soft spot in his heart for Virginia City. Although the main street is a bit too much of a tourist trap for me, the rest of the town, back up on the hill, is fascinating. As a matter of fact, even the honky-tonk atmosphere of the main street is not out of character. These old towns were pretty wild.

Another old town is Bodie near the California-Nevada boundary just north of Mono Lake. Now a State Historic park, this genuine old mining town is maintained by the State of California with much help from The Friends of Bodie. Still another State Historic Park is Columbia, on the southern end of Highway 49. Here the shops have been restored and are in use, with merchants encouraged to sell goods similar to those sold to miners in the heyday of the local mines. Columbia is more nearly in its original condition than any other Sierra foothill town. You can take the kids on a ride aboard a genuine Wells Fargo stagecoach with different prices for riding inside or riding "shotgun" next to the driver. Kids love it. Or you can take a tour through the Hidden Treasure Gold Mine, with over 700 feet of tunnel burrowing through a rich lode. It's hard-rock mining, not placer mining as you'll be doing, but it's exciting to remember that over $50,000,000 in gold passed through Columbia during the ten years of the gold rush days.

Oatman, Arizona, just southwest of Kingman in the northwest corner of the state, has an interesting attraction and an unusual history. The attraction is crowds of desert burros, whose ancestors were abandoned maybe a hundred years ago when their prospector/owner died or moved on. The burros thrived; they (or their descendants) now live in the surrounding hills and crowd into town on weekends-- when the tourist will be there--looking for a handout of burro food, obligingly sold to the tourists by the shopkeepers on the mainstreet.

The recent history of Oatman has given it the nickname of the town that died twice. It and its neighbor Goldroad, were thriving gold towns in the early part of this century. Goldroad depended on its mining almost entirely; Oatman, being on old US route 66, did a good business in tourism as well as the mining industry.

But then came the War--World War II. In an all-out effort to win, many industries were declared non-essential and shut down. Gold mining was one of them. Oatman died; no mining meant no income. However, the inhabitants had heart, and they were in a strategic location--on route 66 just before it took off into the desert of Southern California. So the town turned from mining to tourism and did a good business pumping gas and frying burgers.

Then the last straw was when the Great White Fathers in Washington, who had been planning a replacement for bumpy old 66, decreed that the new highway, Interstate 40, should go over the mountains well to the south of Oatman. The town died a second time, as grass sprouted in the cracks of the old highway.

But there was still life in the tough old town. As the old timers sat on their porches and watched the burros mooching for handouts, it dawned on then that they, too, were a part of history. So they renovated the old saloon and the hash house and the hotel where Clark Gable and Carole Lombard honeymooned, and now they're doing quite well thank you. And rightly so; the old place is well worth seeing.

Oatman's neighbor Goldroad is off the main highway on a side road. Today there is nothing left there but old ruins--photographic ruins, to be sure, and lots of interesting scenery. It's a quiet place; even the wind seems a little muted. Personally, the quiet, almost haunted places like Goldroad impress me more than the restored towns.

Colorado also has many ghost towns worth exploring. The *Atlases of Colorado Ghost Towns, Vol. I & II*, published by Cache Press of Deming, New Mexico are an excellent resource with complete listings.

Many books are available showing the location of the old ghost towns of the West, with some pictures and descriptions of the more important ones. Books such as *Arizona Treasure Hunter's Ghost Town Guide, California Ghost Town Trails, Death Valley Ghost Towns* (Vol. 1 & 2), *Nevada Ghost Town Trails* and the *Southwestern Ghost Town Atlas* are well worth buying; they can be had at most bookstores and also in rock shops.

The abandoned towns vary from a single remaining shack to a collection of recognizable buildings. They are often almost buried in the brush and vegetation of the region and are usually well off the beaten track. These silent, almost haunted places impress me more than the restored towns and I find it a fascinating hobby to "collect" ghost towns. They're very photogenic, taken from almost any angle. Black and white film is often more effective than color, in bringing out the stark desolation of the places.

If you want to go out exploring for ghost towns, here are some suggestions about where to try. In California, State Route 49 is dotted with mementos of the gold rush days. It's also a very scenic route. Take your time and turn off on the little side roads now and then, to see what you can find. Some of the sure bets are Angel's Camp, Altaville, Sutter's Creek, Amador, Chinese Camp, Douglas Flat, and Fiddletown, but there are dozens of little settlements off the main road near these places.

A second region to explore, in fall, winter, or spring is the Panamint region on the western edge of Death Valley. Go east and then north from China Lake, through Trona, and then about 30 miles north. Then start looking and explore the dirt roads. There were two main settlements here in the old days – Panamint and Ballarat. Both are genuine ghost towns and practically unoccupied. In the same desert area is Garlock, a collection of a dozen or so ruins on a dirt road running from Randsburg, south of China Lake, over to State Route 14. Randsburg itself is interesting. It's not really a ghost town, since it is occupied, but the region around the town is dotted with old mine shafts and the remains of a gold rush.

Another region to explore is the backbone of the Sierra Nevada – the eastern flank. Here you will find Bodie and Masonic, both near Bridgeport on U.S. 395 north of Mono Lake. Both towns are on dirt roads, the road to Bodie going east just south of Bridgeport, and the one to Masonic branching off of State Route 182.

There are some interesting places in Arizona, too. Try the eastern side of the Colorado River – Ehrenburg, La Paz and Quartzsite are within 15 miles of Blythe on Interstate 10. Also try near the little town of Congress, just north of Wickenburg on U.S. 89. Congress itself has been rebuilt, but the ruins of the old town are just 3 or 4 miles north of it on a dirt road. Another region in Arizona is around Prescott, on Lynx Creek, where the first gold in Arizona was discovered.

Ghost Coins and Bottles

Finding an abandoned ghost town off the beaten track is an exciting experience. The old ruins are also the ideal place to use your metal detector. The original occupants very often left souvenirs, coins and jewelry behind. Search the trash dumps and hunt all around inside the old ruins. Don't forget the walls. People often hide things in walls and then forget about them.

Any coins you find are usually old ones, and often valuable. One issue of a Los Angeles paper told of a man who found an old dime worth $65, and ten Indian head pennies in an abandoned ruin. A companion found a gold chain with solid gold nuggets fastened to the links.

There are often old bottles around old ruins, but for some reason most people ignore them. That's a mistake, because old bottles are well worth collecting. Many of them have mottled color from the sunlight – this enhances their value. Almost any bottle from the last century will bring $5, and rare ones may be worth ten times that. Look for bottles with embossed lettering on them, especially if they once held bitters.

One of the best clues to a bottle's age is the seam, or mold line, on the side of a bottle. You can feel it if you run your hand carefully along the bottle.   If the seam runs all the way up to the top of the bottle, forget it. It's modern. If the seam ends at the shoulder (the bottle's curves to the neck) or even halfway between the shoulder and the top – keep it and get it appraised by a good collector. Also, look for poor workmanship in the bottle – uneven glass thickness, neck set on a slight angle, bubbles in the glass, etc. These are clues that the bottle was made by hand, instead of on a modern machine.

Be careful when you explore these places. Wells and old mine shafts are dangerous traps, and snakes and spiders have sometimes taken possession of the ruins. Just make plenty of noise as you tramp around, and the snakes will get out of your way. Snakes don't like people any more than people like snakes. But if they're surprised, they will strike. Wouldn't you?

Lost Mines

Lost mines are even more elusive and fascinating than ghost towns. Many of the yarns are myths or gross exaggerations, of course, but there is no question that a great many profitable mines were abandoned by their owners, who disappeared under mysterious circumstances. On the other hand, sometimes the circumstances were quite clear. The lost Lee Mine, a fabulous ledge of gold-bearing quartz in the Bullion

Mountains near Barstow, was registered by the finder in 1890. According to the record, the gentleman was then shot through the heart in San Bernardino. The murderer was never found; neither was the mine, though ex-governor Waterman, a miner, offered a huge reward for information as to its whereabouts.

In the same general location, an old prospector named Hermit John entered Amboy, California, with a sack of ore – very rich, and different from the ore being produced by the nearby Virginia Dale Mine. Over a bottle of rot-gut whiskey, Hermit John let slip the fact that the ore had come from the Sheephole Mountains, northeast of Dale Dry Lake and southwest of Cadiz Dry Lake. Hermit John finished his drink, went back to the desert, and was never heard from again.

Some of the lost mines, such as the wonderful Whiteman cement mine and the Lost Dutchman Mine, have been the subject of books and movies. A fascinating book, *Nevada Lost Mines and Buried Treasures* by Douglas McDonald, describes the stories of over 70 lost mines in Nevada and *Dig Here!* by Thomas Penfield gives probable value, general location and an estimate of authenticity for 65 lost mines in the Arizona region. Both books are essential to lost mine hunters, and both are entertaining reading for stay-at-homes.

One of the famous lost mines of the Mother Lode country, the Whiteman "cement" mine, is described in priceless fashion by Mark Twain in *Roughing It* (chapter 37). It gives a good idea of how mines got lost.

In brief, three young German brothers, fleeing from an Indian attack, came to a mountain gorge somewhere in the neighborhood of Mono Lake, on the eastern flank of the Sierra Nevada. They staggered, starving and disoriented, through the gorge and collapsed on a ledge to rest. One of them noticed that the ledge looked like a vein of cement shot full of lumps of dull yellow metal. They saw it was gold – a fortune! Each of the brothers loaded himself with about twenty-five pounds of it and headed westward again. Troubles struck them down. One of the brothers fell and broke his leg and the other two were obliged to leave him and go on. A second, worn out and starving, gave up and fell by the trail to die. But the third finally reached the gold settlements of California, exhausted, sick and out of his mind. He had thrown away all of his cement except for a few fragments, but these were enough to set off a frenzy of excitement.

The brother had his fill of gold, and settled down to work on a farm for wages. But he gave his map to a man named Whiteman, and described the cement region as well as he could, and (to quote Mark Twain) "...thus transferred the curse to that gentleman, for when I had my one accidental glimpse of Mr. W. in Esmeralda, he had been hunting for the lost mine, in hunger and thirst, poverty and sickness, for twelve or thirteen years...I saw a piece of cement as large as my fist which was said to have been given to Whiteman by the young German, and it was of a seductive nature. Lumps of gold were as thick in it as raisins in a slice of fruit cake. The privilege of working such a mine one week would be sufficient for a man of reasonable desires."

One more yarn about lost mines; this time a modern one from the pages of the *Mohave Daily Miner* of August, 1986. A man who has been claiming to have found the Lost Dutchman Mine has been sentenced to refund nearly half a million dollars to his backers. The man, Robert S. Jacob of Chandler, Arizona, has been claiming since 1965 that he found the mine; over the years he convinced people to invest millions of dollars in a venture to bring the gold out of the Superstition Mountains. He was charged with fraud in December of 1985 and pleaded guilty.

A footnote: the prosecuting attorney, Assistant Attorney General Barnett Lofstein, points out that the guilty plea does not indicate that the mine does not exist. "We don't know whether it existed or not, just that Mr. Jacob didn't find it," said Lofstein.

Well...We can't all find a lost mine, and we can't all come home to the local saloon and plunk down a heavy bag of gold dust. But we can have a lot more fun than the old-timers did. There's a lot more to life than gold. There's fresh air, and blue skies, and the joy of feeling tired muscles. We can all seek out these pleasures on the weekends, in this great free country of ours.

So whether you find gold or not, you'll get a rich reward. Good luck to you! Good prospecting!